Joanne Rankin

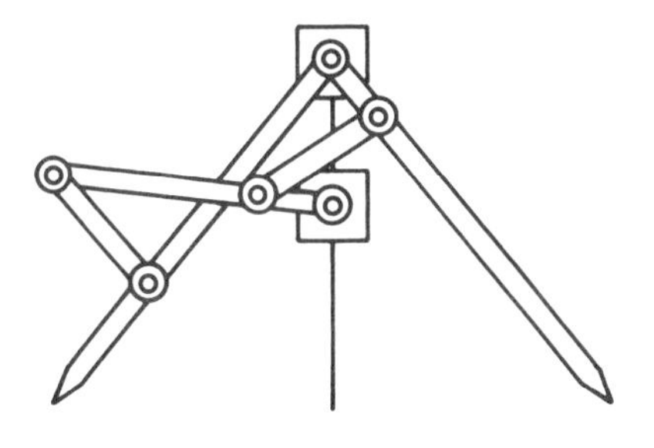

HOW TO DRAW A STRAIGHT LINE

CLASSICS
IN MATHEMATICS EDUCATION

Volume 1: *The Pythagorean Proposition*
by Elisha Scott Loomis

Volume 2: *Number Stories of Long Ago*
by David Eugene Smith

Volume 3: *The Trisection Problem*
by Robert C. Yates

Volume 4: *Curves and Their Properties*
by Robert C. Yates

Volume 5: *A Rhythmic Approach to Mathematics*
by Edith L. Somervell

Volume 6: *How to Draw a Straight Line: A Lecture on Linkages*
by A. B. Kempe

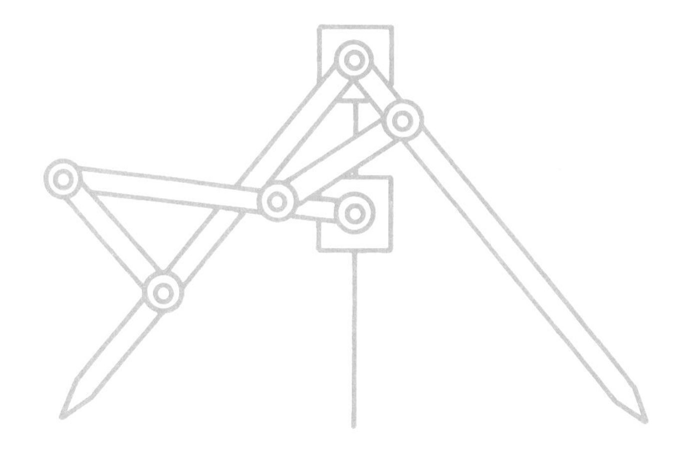

HOW TO DRAW A STRAIGHT LINE

A. B. Kempe

THE NATIONAL COUNCIL OF TEACHERS OF MATHEMATICS

1906 Association Drive, Reston, Virginia 22091

Library of Congress Cataloging in Publication Data

Kempe, Alfred Bray, 1849-
How to draw a straight line.

(Classics in mathematics education; v. 6)
Reprint of the ed. published by the Pentagon, Albion, Mich., which was a reprint of the 1877 ed. published by Macmillan, London, in series: Nature series.
Includes bibliographical references.
1. Straight-line mechanisms. I. Title.
II. Series: Nature series.
TJ181.8.K45 516'.183 77-6669
ISBN 0-87353-120-5

Printed in the United States of America

1977

FOREWORD

Some mathematical works of considerable vintage have a timeless quality about them. Like classics in any field, they still bring joy and guidance to the reader. Books of this kind, when they concern fundamental principles and properties of school mathematics and are no longer readily available, are being sought by the National Council of Teachers of Mathematics, which began publishing a series of such classics in 1968. The present title is the sixth volume in the series.

How to Draw a Straight Line: A Lecture on Linkages was originally published in London by Macmillan and Company in 1877. The author was a member of the Council of the London Mathematical Society and a scholar of Trinity College, Cambridge. The book was reprinted in the United States in the *Pentagon,* a journal of Kappa Mu Epsilon, national honorary mathematics society. Because no copy of the original edition could be found, the type has been reset for this edition. The illustrations are those that appeared in the *Pentagon* reprint, not the original art prepared by the author's brother, H. R. Kempe. No attempt has been made to edit or modernize the text in any way. To do so would surely detract from, rather than add to, its value.

NOTICE

This Lecture was one of the series delivered to science teachers last summer in connection with the Loan Collection of Scientific Apparatus. I have taken the opportunity afforded by its publication to slightly enlarge it and to add several notes. For the illustrations I am indebted to my brother, Mr. H. R. Kempe, without whose able and indefatigable co-operation in drawing them and in constructing the models furnished by me to the Loan Collection I could hardly have undertaken the delivery of the Lecture, and still less its publication.

7, Crown Office Row, Temple,
January 16th, 1877.

The great geometrician Euclid, before demonstrating to us the various propositions contained in his *Elements of Geometry*, requires that we should be able to effect certain processes. These *Postulates*, as the requirements are termed, may roughly be said to demand that we should be able to describe straight lines and circles. And so great is the veneration that is paid to this master-geometrician, that there are many who would refuse the designation of "geometrical" to a demonstration which requires any other construction than can be effected by straight lines and circles. Hence many problems—such as, for example, the trisection of an angle—which can readily be effected by employing other simple means, are said to have no geometrical solution, since they cannot be solved by straight lines and circles only.

It becomes then interesting to inquire how we can effect these preliminary requirements, how we can describe these circles and these straight lines, with as much accuracy as the physical circumstances of the problems will admit of.

As regards the circle we encounter no difficulty. Taking Euclid's definition, and assuming, as of course we must, that our surface on which we wish to describe the circle is a plane, (1)[1] we see that we have only to make our tracing-point preserve a distance from the given centre of the circle constant and equal to the required radius. This can readily be effected by taking a flat piece of any form, such as the piece of cardboard I have here, and passing a pivot which is fixed to the given surface at the given centre through a hole in the piece, and a tracer or pencil through another hole in it whose distance from the first is equal to the given radius; we shall then, by moving the pencil, be able, even with this rude apparatus, to describe a circle with considerable accuracy and ease; and when we come to employ very small holes and pivots, or even large ones, turned with all that marvelous truth which the lathe affords, we shall get a result unequalled perhaps among mechanical apparatus for the smoothness and accuracy of its movement. The apparatus I have just described is of course nothing but a simple form of a pair of compasses, and it is usual to say that the third Postulate postulates the compasses.

But the straight line, how are we going to describe that? Euclid defines it as "lying evenly

[1] These figures refer to notes at the end of the lecture.

between its extreme points." This does not help us much. Our textbooks say that the first and second Postulates postulate a ruler (2). But surely that is begging the question. If we are to draw a straight line with a ruler, the ruler must itself have a straight edge; and how are we going to make the edge straight? We come back to our starting-point.

Now I wish you clearly to understand the difference between the method I just now employed for describing a circle, and the ruler method of describing a straight line. If I applied the ruler method to the description of a circle, I should take a circular lamina, such as a penny, and trace my circle by passing the pencil around the edge, and I should have the same difficulty that I had with the straight-edge, for I should first have to make the lamina itself circular. But the other method I employed involves no begging the question. I do not first assume that I have a circle and then use it to trace one, but simply require that the distance between two points shall be invariable. I am of course aware that we do employ circles in our simple compass, the pivot and the hole in the moving piece which it fits are such; but they are used not because they are the curves we want to describe (they are not so, but are of a different size), as is the case with the straight-edge, but because, through the impossibility of constructing pivots or holes of no finite dimensions, we are forced to adopt the best substitute we can for making one point in the moving piece remain at the same spot. If we employ a very small pivot and hole, though they be not truly circular, the error in the description of a circle of moderate dimensions will be practically infinitesimal, not perhaps varying beyond the width of the thinnest line which the tracer can be made to describe; and even when we employ large pivots and holes we shall get results as accurate, because those pivots and holes may be made by the employment of very small ones in the machine which makes them.

It appears then, that although we have an easy and accurate method of describing a circle, we have at first sight no corresponding means of describing a straight line and there would seem to be a substantial difficulty in producing what mathematicians call the simplest curve, so that the question how to get over that difficulty becomes one of a decided theoretical interest.

Nor is the interest theoretical only, for the question is one of direct importance to the practical mechanician. In a large number of machines and scientific apparatus it is requisite that some point or points should move accurately in a straight line with as little friction as possible. If the ruler principle is adopted, and the point is kept in its path by guides, we have, besides the initial difficulty of making the guides truly straight, the wear and tear produced by the friction of the sliding surfaces, and the deformation produced by changes of temperature and varying strains. It becomes therefore of real consequence to obtain, if possible, some method which shall not involve these objectionable features, but possess the accuracy and ease of movement which characterize our circle-producing apparatus.

Turning to that apparatus, we notice that all that is requisite to draw with accuracy a circle of any given radius is to have the distance between the pivot and the tracer properly determined, and if I pivot a second "piece" to the fixed surface at a second point having a tracer as the first piece has, by properly determining the distance between the second tracer and pivot, I can describe a second circle whose radius bears any proportion I please to that of the first circle. Now, removing the tracers, let me pivot a third piece to these two *radial* pieces, as I may call them, at the points where the tracers were, and let me fix a tracer at any point on this third or *traversing* piece. You will at once see that if the radial pieces were big enough the tracer would describe circles or portions of circles on *them*, though they are in motion, with the same ease and accuracy as in the case of the simple circle-drawing apparatus; the tracer will not however describe a circle on the *fixed* surface, but a complicated curve.

This curve will, however, be described with all the ease and accuracy of movement with which the circles were described, and if I wish to reproduce in a second apparatus the curves which I produce with this, I have only to get the distances between the pivots and tracers accurately the same in both cases, and the curves will also be

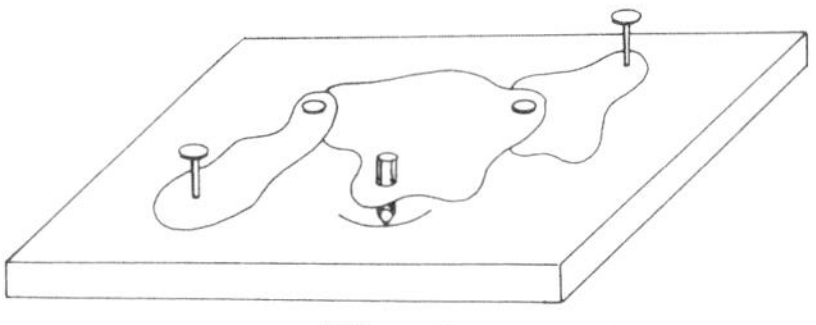

Fig. 1.

accurately the same. I could of course go on adding fresh pieces *ad libitum*, and I should get points on the structure produced, describing in general very complicated curves, but with the same results as to accuracy and smoothness, *the reproduction of any particular curve depending solely on the correct determination of a certain definite number of distances.*

These systems, built up of pieces pointed or pivoted together, and turning about pivots attached to a fixed base, so that the various points on the pieces all describe definite curves, I shall term "link-motions," the pieces being termed "links." As, however, it sometimes facilitates the consideration of the properties of these structures to regard them apart from the base to which they are pivoted, the word "linkage" is employed to denote any combination of pieces pivoted together. When such a combination is pivoted in any way to a fixed base, the motion of points on it not being necessarily confined to fixed paths, the link-structure is called a "linkwork:" a "linkwork" in which the motion of every point is in some definite path being, as before stated, termed a "link-motion." I shall only add to these expressions two more: the point of a link-motion which describes any curve is called a "graph," the curve being called a "gram" (3).

The consideration of the various properties of these "linkages" has occupied much attention of late years among mathematicians, and is a subject of much complexity and difficulty. With the purely mathematical side of the question I do not, however, propose to deal today, as we shall have quite enough to do if we confine our attention to the practical results which mathematicians have obtained, and which I believe only mathematicians could have obtained. That these results are valuable cannot I think be doubted, though it may well be that their great beauty has led some to attribute to them an importance which they do not really possess; and it may be that fifty years ago they would have had a value which, through the great improvements that modern mechanicians have effected in the production of true planes, rulers and other exact mechanical structures, cannot now be ascribed to them. But linkages have not at present, I think, been sufficiently put before the mechanician to enable us to say what value should really be set upon them.

The practical results obtained by the use of linkages are but few in number, and are closely connected with the problem of "straight-line motion," having in fact been discovered during the investigation of that problem, and I shall be naturally led to consider them if I make "straight-line motion" the backbone of my lecture. Before, however, plunging into the midst of these linkages it will be useful to know how we can practically construct such models as we require; and here is one of the great advantages of our subject—we can get our results visibly before us so very easily. Pins for fixed pivots, cards for links, string or cotton for the other pivots, and a dining-room table, or a drawing-board if the former be thought objectionable, for a fixed base, are all we require.

If something more artistic be preferred, the plan adopted in the models exhibited by me in the Loan Collection can be employed. The models were constructed by my brother, Mr. H. R. Kempe, in the following way. The bases are thin deal boards painted black; the links are neatly shaped out of thick cardboard (it is hard work making them, you have to sharpen your knife about every ten minutes, as the cardboard turns the edge very rapidly) ; the pivots are little rivets made of catgut, the heads being formed by pressing the face of a heated steel chisel on the ends of the gut after it is passed through the holes in the links; this gives a very firm and smoothly-working joint. More durable links may be made of tinplate; the pivot-holes must in this case be punched, and the eyelets used by bootmakers for laced boots employed as pivots; you can get the proper tools at a trifling expense at any large tool shop.

Now, as I have said, the curves described by the various points on these link-motions are in general very complex. But they are not necessarily so. By properly choosing the distances at our disposal we can make them very simple. But can we go to the fullest extent of simplicity and get a point on one of them moving accurately in a straight line? That is what we are going to investigate.

To solve the problem with our single link is clearly impossible: all the points on it describe circles. We must therefore go to the next simple case—our three-link motion. In this case you will see that we have at our disposal the distance between the fixed pivots, the distances between the pivots on the radial links, the distance between the pivots on the traversing link, and the distances of the tracer from those pivots; in all six different distances. Can we choose those distances so that our tracing-point shall move in a straight line?

The first person who investigated this was the great man James Watt. "Watt's Parallel Motion" (4), invented in 1784, is well known to every engineer, and is employed in nearly every beam-engine. The apparatus, reduced to its simplest form, is shown in Fig. 2.

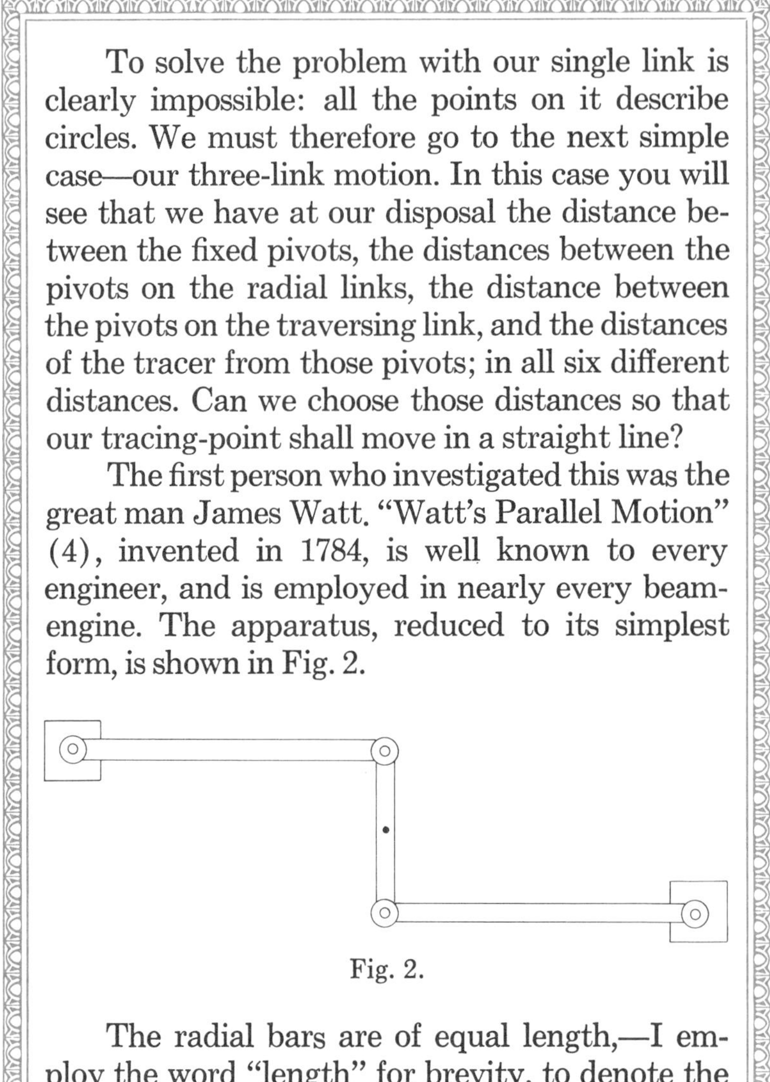

Fig. 2.

The radial bars are of equal length,—I employ the word "length" for brevity, to denote the distance between the pivots; the links of course may be of any length or shape,—and the distance between the pivots or the traversing link is such that when the radial bars are parallel the line joining those pivots is perpendicular to the radial bars. The tracing-point is situated half-way between the pivots on the traversing piece. The curve described by the tracer is, if the apparatus does not deviate much from its mean position, approximately a straight line. The reason of this is that the circles described by the extremities of the radial bars have their concavities turned in opposite directions, and the tracer being half-way between, describes a curve which is concave neither one way nor the other, and is therefore a straight line. The curve is not, however, accurately straight, for if I allow the tracer to describe the whole path it is capable of describing, it will, when it gets some distance from its mean position, deviate considerably from the straight line, and will be found to describe a figure 8, the portions at the crossing being nearly straight. We know that they are not quite straight, because it is impossible to have such a curve partly straight and partly curved.

For many purposes the straight line described by Watt's apparatus is sufficiently accurate, if we require an exact one it will, of course, not do, and we must try again. Now it is capable of proof that it is impossible to solve the problem with three moving links; closer approximations to the truth than that given by Watt can be obtained, but still not actual truth.

I have here some examples of these closer approximations. The first of these, shown in Fig. 3, is due to Richard Roberts of Manchester.

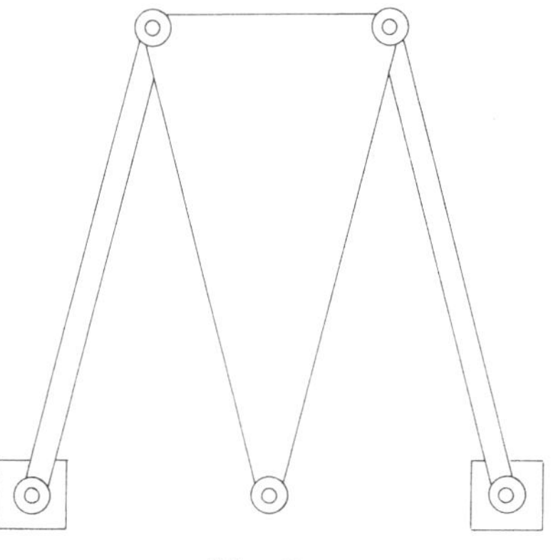

Fig. 3.

The radial bars are of equal length, the distance between the fixed pivots is twice that of the pivots on the traversing piece, and the tracer is situate on the traversing piece, at a distance from the pivots on it equal to the lengths of the radial bars. The tracer in consequence coincides with the straight line joining the fixed pivots at those pivots and half-way between them. It does not, however, coincide at any other points, but deviates very slightly between the fixed pivots. The path described by the tracer when it passes the pivots altogether deviates from the straight line.

The other apparatus was invented by Professor Tchebicheff of St. Petersburg. It is shown in Fig. 4. The radial bars are equal in length, being each in my little model five inches long.

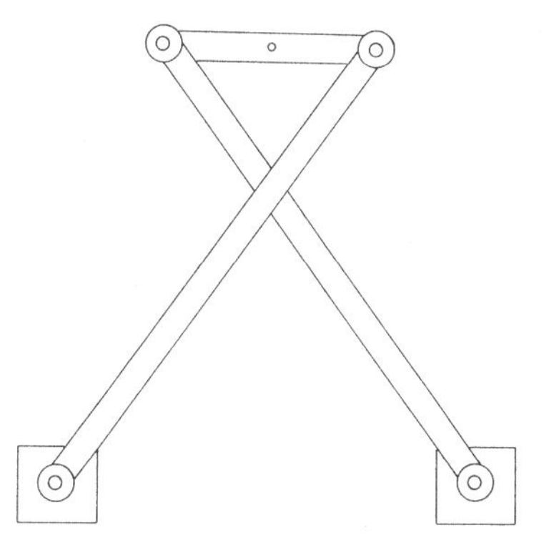

Fig. 4.

The distance between the fixed pivots must then be four inches, and the distance between the pivots of the traversing bar two inches. The tracer is taken half-way between these last. If now we draw a straight line—I had forgotten that we cannot do that yet, well, if we draw a straight line, popularly so called—through the tracer in its mean position, as shown in the figure, parallel to that forming the fixed pivots, it will be found that the tracer will coincide with that line at the points where verticals through the fixed pivots cut it as well as at the mean position, but, as in the case of Robert's parallel motion, it coincides nowhere else, though its deviation is very small as long as it remains between the verticals.

We have failed then with three links, and we must go on to the next case, a five-link motion—for you will observe that we must have an odd number of links if we want an apparatus describing definite curves. Can we solve the problem with five? Well, we can; but this was not the first accurate parallel motion discovered, and we must give the first inventor his due (although he did not find the simplest way) and proceed in a strict chronological order.

In 1864, eighty years after Watt's discovery, the problem was first solved by M. Peaucellier, an officer of Engineers in the French army. His discovery was not at first estimated at its true value, fell almost into oblivion, and was rediscovered by a Russian student named Lipkin, who got a substantial reward from the Russian Government for his supposed originality. However, M. Peaucellier's merit has at last been recognized, and he has been awarded the great mechanical prize of the Institute of France, the "Prix Montyon."

M. Peaucellier's apparatus is shown in Fig. 5. It has, as you see, seven pieces or links. There are first of all two long links of equal length. These are both pivoted at the same fixed point; their other extremities are pivoted to opposite angles of a rhombus composed of four equal shorter links.

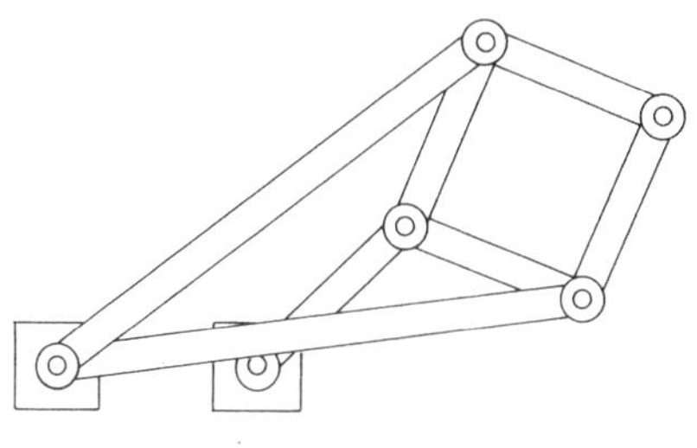

Fig. 5.

The portion of the apparatus I have thus far described, considered apart from the fixed base, is a linkage termed a "Peaucellier cell." We then take an *extra* link, and pivot it to a fixed point whose distance from the first fixed point, that to which the cell is pivoted, is the same as the length of the extra link; the other end of the extra link is then pivoted to one of the free angles of the rhombus; the other free angle of the rhombus has a pencil at its pivot. That pencil will accurately describe a straight line.

I must now indulge in a little simple geometry. It is absolutely necessary that I should do so in order that you may understand the principle of our apparatus.

In Fig. 6, QC is the extra link pivoted to the fixed point, Q, the other pivot on it C, describing the circle OCR. The straight lines PM and PM' are supposed to be perpendicular to $MRQOM'$.

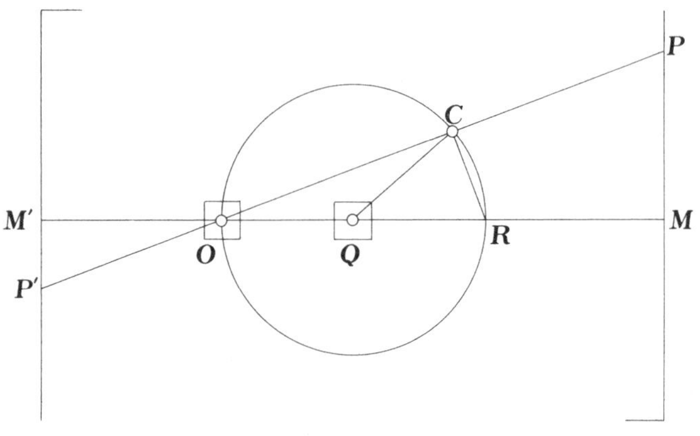

Fig. 6.

Now the angle OCR, being the angle in a semicircle, is a right angle. Therefore the triangles OCR, OMP are similar. Therefore,

$$OC : OR = OM : OP.$$

Therefore,

$$OC \cdot OP = OM \cdot OR,$$

wherever C may be on the circle. That is, since OM and OR are both constant, if while C moves in a circle P moves so that O, C, P are always in the same straight line, and so that $OC \cdot OP$ is always constant; then P will describe the straight line PM perpendicular to the line OQ.

It is also clear that if we take the point P' on the other side of O, and if $OC \cdot OP'$ is constant P' will describe the straight line $P'M'$. This will be seen presently to be important.

Now turning to Fig. 7, which is a skeleton drawing of the Peaucellier cell, we see that from

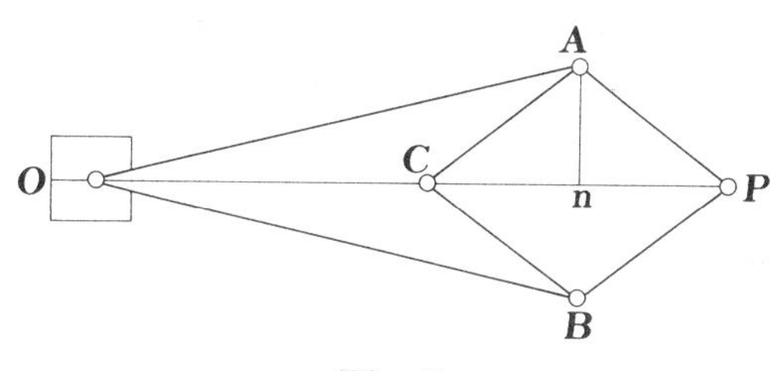

Fig. 7.

the symmetry of the construction of the cell, O, C, P, all lie in the same straight line, and if the straight line An be drawn perpendicular to CP—it must still be an imaginary one, as we have not proved yet that our apparatus does draw a straight line—Cn is equal to nP. Now

$$OA^2 = On^2 + An^2$$

$$AP^2 = Pn^2 + An^2$$

therefore,

$$\begin{aligned} OA^2 - AP^2 &= On^2 - Pn^2 \\ &= (On - Pn)\ (On + Pn) \\ &= OC \cdot OP. \end{aligned}$$

Thus since OA and AP are both constant $OC \cdot OP$ is always constant, however far or near C and P may be to O. If then the pivot O be fixed to the point O in Fig. 6, and the pivot C be made to describe the circle in the figure by being pivoted to the end of the extra link, the pivot P will satisfy all the conditions necessary to make it move in a straight line, and if a pencil be fixed at P it will draw a straight line. The distance of the line from the fixed pivots will of course depend on the magnitude of the quantity $OA^2 - OP^2$ which may be varied at pleasure.

I hope you clearly understand the two elements composing the apparatus, the extra link and the cell, and the part each plays, as I now wish to describe to you some modifications of the cell. The extra link will remain the same as before, and it is only the cell which will undergo alteration.

If I take the two linkages in Fig. 8, which are known as the "kite" and the "spear-head," and

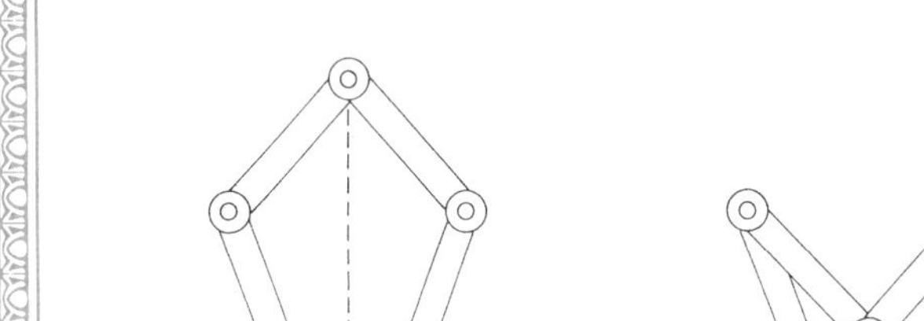

Fig. 8.

place one on the other so that the long links of the one coincide with those of the other, and then amalgamate the coincident long links together, we shall get the original cell of Figs. 5 and 7. If then we keep the angles between the long links, or that between the short links, the same in the "kite" and "spear-head," we see that the height of the "kite" multiplied by that of the "spear-head" is constant.

Let us now, instead of amalgamating the long links of the two linkages, amalgamate the short ones. We then get the linkage of Fig. 9; and if the pivot where the short links meet is fixed, and one of the other free pivots be made to move in the circle of Fig. 6 by the extra link, the other will describe, not the straight line PM, but the straight line $P'M'$. In this form, which is a very compact one, the motion has been applied in a beautiful manner to the air engines which are employed to ventilate the Houses of Parliament. The ease of working and absence of friction and noise is remarkable. The engines were constructed and the Peaucellier apparatus adapted to them

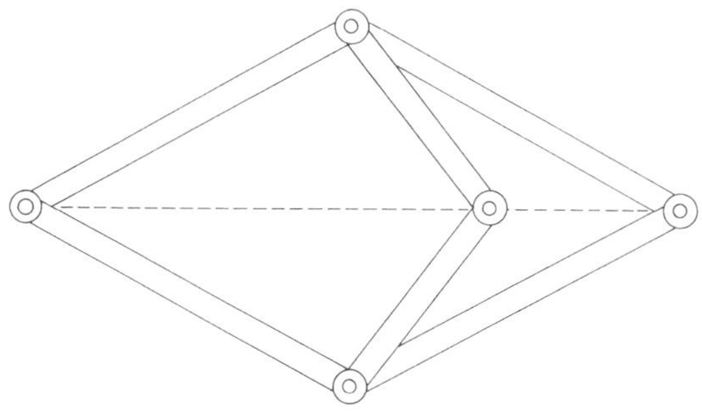

Fig. 9.

by Mr. Prim, the engineer to the Houses, by whose courtesy I have been enabled to see them, and I can assure you that they are well worth a visit.

Another modification of the cell is shown in Fig. 10. If instead of employing a "kite" and "spear-head" of the same dimensions, I take the same "kite" as before, but use a "spear-head" of half the size of the former one, the angles being however kept the same, the product of the heights of the two figures will be half what it was before,

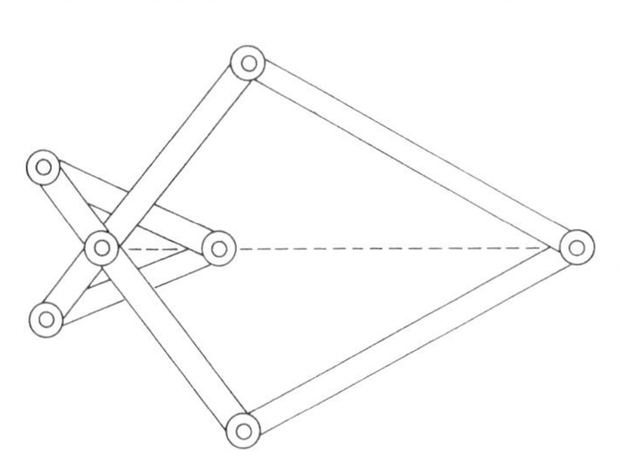

Fig. 10.

but still constant. Now instead of superimposing the links of one figure on the other, it will be seen that in Fig. 10 I fasten the shorter links of each figure together, end to end. Then, as in the former cases, if I fix the pivot at the point where the links are fixed together, I get a cell which may be used, by the employment of an extra link, to describe a straight line. A model employing this form of cell is exhibited in the Loan Collection by the Conservatoire des Arts et Métiers of Paris, and is of exquisite workmanship; the pencil seems to swim along the straight line.

Mr. Peaucellier's discovery was introduced into England by Professor Sylvester in a lecture he delivered at the Royal Institution in January, 1874 (5), which excited very great interest and was the commencement of the consideration of the subject of linkages in this country.

In August of the same year Mr. Hart of Woolwich Academy read a paper at the British Association meeting (6), in which he showed that M. Peaucellier's cell could be replaced by an apparatus containing only four links instead of six. The new linkage is arrived at thus.

If to the ordinary Peaucellier cell I add two fresh links of the same length as the long ones I get the double, or rather quadruple cell, for it may be used in four different ways, shown in Fig. 11. Now Mr. Hart found that if he took an ordinary parallelogrammatic linkwork, in which the

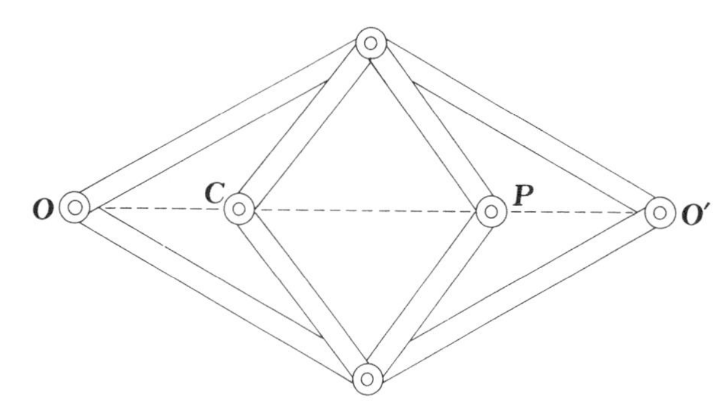

Fig. 11.

adjacent sides are unequal, and crossed the links so as to form what is called a contra-parallelogram, Fig. 12, and then took four points on the four links dividing the distances between the pivots in the same proportion, those four points

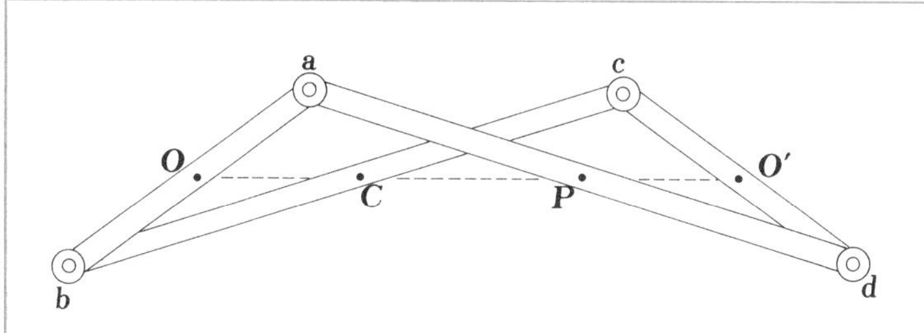

Fig. 12.

had exactly the same properties as the four points of the double cell. That the four points always lie in a straight line is seen thus: considering the triangle abd, since $aO : Ob = aP : Pd$ therefore OP is parallel to bd, and the perpendicular distance between the parallels is to the height of the triangle abd as Ob is to ab; the same reasoning applies to the straight line CO', and since $ab : Ob = cd : O'd$ and the heights of the triangles abd, cbd, are clearly the same, therefore the distances of OP and $O'C$ from bd are the same, and $OCPO'$ lie in the same straight line.

That the product $OC \cdot OP$ is constant appears at once when it is seen that ObC is half a "spear-head" and OaP half a "kite;" similarly it may be shown that $O'P \cdot O'C$ is constant, as also $OC \cdot CO'$ and $OP \cdot PO'$. Employing then the Hart's cell as we employed Peaucellier's, we get a five-link straight line motion. A model of this is exhibited in the Loan Collection by M. Breguet.

I now wish to call your attention to an extension of Mr. Hart's apparatus, which was discovered simultaneously by Professor Sylvester and myself. In Mr. Hart's apparatus we were only concerned with bars and points on those bars, but in the apparatus I wish to bring before you we have pieces instead of bars. I think it will be more interesting if I lead up to this apparatus by detailing to you its history, especially as I shall thereby be enabled to bring before you another very elegant and very important linkage—the discovery of Professor Sylvester.

When considering the problem presented by the ordinary three-*bar* motion consisting of two radial bars and a traversing bar, it occurred to me—I do not know how or why, it is often very difficult to go back and find whence one's ideas originate—to consider the relation between the curves described by the points on the traversing bar in any given three-bar motion, and those described by the points on a similar three-bar motion, but in which the traversing bar and one of the radial bars had been made to change places. The proposition was no sooner stated than the solution became obvious; the curves were precisely similar. In Fig. 13 let CD and BA be the two radial bars turning about the fixed centers C and B, and let DA be the traversing bar, and let

P be any point on it describing a curve depending on the lengths of AB, BC, CD, and DA. Now add to the three-bar motion the bars CE and EAP', CE being equal to DA, and EA equal to CD.

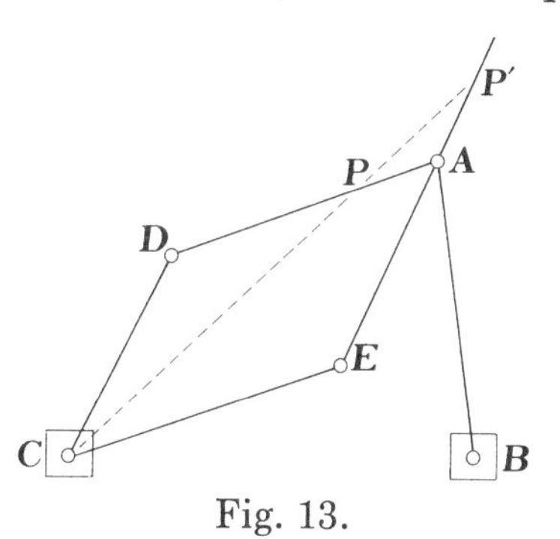

Fig. 13.

$CDAE$ is then a parellogram, and if an imaginary line CPP' be drawn, cutting EA produced in P', it will at once be seen that P' is a fixed point on EA produced, and CP' bears always a fixed proportion to CP, viz., $CD : CE$. Thus the curve described by P' is precisely the same as that described by P, only it is larger in the proportion $CE : CD$. Thus if we take away the bars CD and DA, we shall get a three-bar linkwork, describing precisely the same curves, only of different magnitude, as our first three-bar motion described, and this new three-bar linkwork is the same as the old with the radial link CD and the traversing link DA interchanged (7).

On my communicating this result to Professor Sylvester, he at once saw that the property was one not confined to the particular case of points lying on the traversing bar, in fact to three-*bar* motion, but was possessed by three-*piece* motion. In Fig. 14 $CDAB$ is a three-bar motion, as in Fig. 13, but the tracing point or "graph" does not lie on the line joining the joints AD, but is anywhere else on a "piece" on which the joints AD lie. Now, as before, add the bar CE, CE being equal to AD, and the piece AEP', making AE

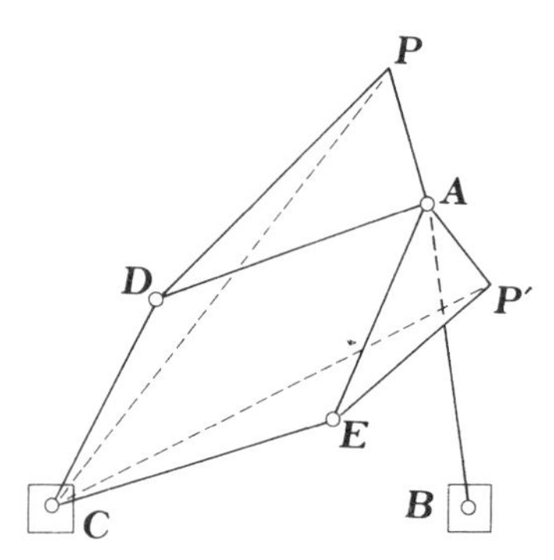

Fig. 14.

equal to CD, and the triangle AEP' similar to the triangle PDA; so that the angles AEP', ADP are equal, and

$$P'E : EA = AD : DP.$$

It follows easily from this—you can work it out

for yourselves without difficulty—that the ratio $P'C : PC$ is constant and the angle PCP' is constant; thus the paths of P and P', or the "grams" described by the "graphs," P and P', are similar, only they are of different sizes, and one is turned through an angle with respect to the other.

Now you will observe that the two proofs I have given are quite independent of the bar AB, which only affects the particular curve described by P and P'. If we get rid of AB, in both cases we shall get in the first figure the ordinary pantagraph, and in the second a beautiful extension of it, called by Professor Sylvester, its inventor, the *Plagiograph* or *Skew Pantagraph*. Like the Pantagraph, it will enlarge or reduce figures, but it will do more, it will turn them through any required angle, for by properly choosing the position of P and P', the ratio of CP to CP' can be made what we please, and also the angle PCP' can be made to have any required value. If the angle PCP' is made equal to 0 or 180°, we get the two forms of the pantagraph now in common use; if it be made to assume successively any value which is a submultiple of 360°, we can, by passing the point P each time over the same pattern make the point P' reproduce it round the fixed center C after the fashion of a kaleidoscope. I think you will see from this that the instrument, which has, as far as I know, never been practically constructed, deserves to be put into the hands of the designer. I give here a picture of a little model of a possible form for the instrument furnished by me to the Loan Collection by request of Professor Sylvester (8).

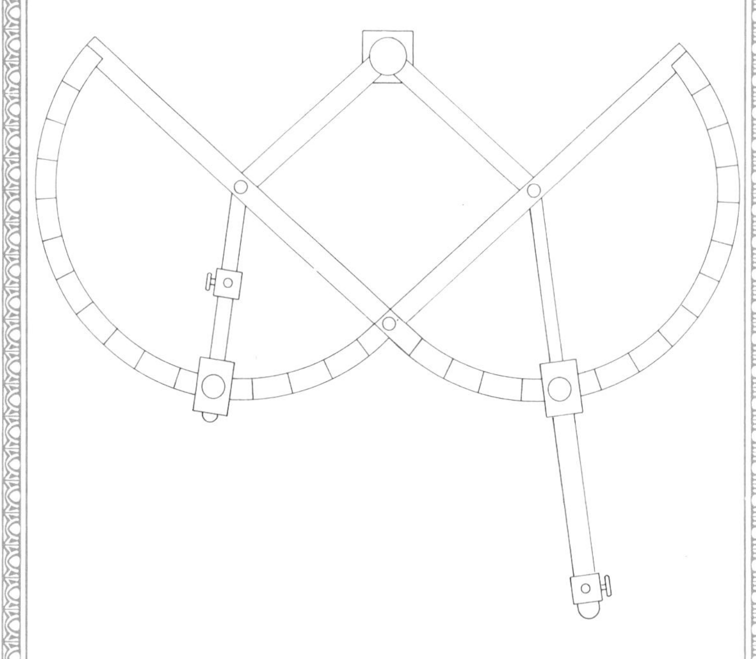

Fig. 15.

After this discovery of Professor Sylvester, it occurred to him and to me simultaneously—our letters announcing our discovery to each other crossing in the post—that the principle of the plagiograph might be extended to Mr. Hart's contra-parallelogram; and this discovery I shall now proceed to explain to you. I shall, however, be more easily able to do so by approaching it in a different manner to that in which I did when I discovered it.

If we take the contra-parallelogram of Mr. Hart, and bend the links at the four points which lie on the same straight line, or *foci* as they are sometimes termed, through the same angle, the four points, instead of lying in the same straight line, will lie at the four angular points of a parallelogram of constant angles,—two the angle that the bars are bent through, and the other two their supplements—and of constant area, so that the product of two adjacent sides is constant.

In Fig. 16 the lettering is preserved as in Fig. 12, so that the way in which the apparatus is formed may be at once seen. The holes are taken in the middle of the links, and the bending is through a right angle. The four holes $OPO'C$ lie at the four corners of a right-angled parallelogram, and the product of any two adjacent sides, as for example $OC \cdot OP$, is constant. It follows that if O be pivoted to the fixed point O in Fig. 6, and C be pivoted to the extremity of the extra link, P

will describe a straight line, not PM, but one inclined to PM at an angle the same as the bars are bent through, *i.e.* a right angle. Thus the straight line will be parallel to the line joining the fixed pivots O and Q. This apparatus, which for simplicity I have described as formed of four

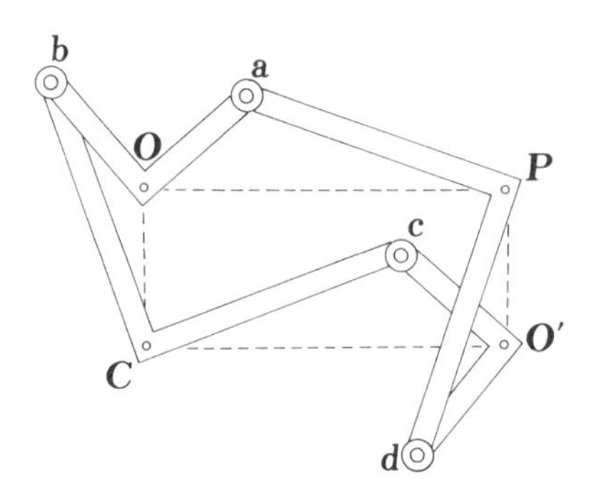

Fig. 16.

straight links which are afterwards bent, is of course strictly speaking formed of four plane links, such as those employed in Fig. 1, on which the various points are taken. This explains the name given to it by Professor Sylvester, the "Quadruplane." Its properties are not difficult to investigate, and when I point out to you that in Fig. 16, as in Fig. 12, Ob, bC form half a "spear-head," and Oa, aP half a "kite," you will very soon get to the bottom of it.

I cannot leave this apparatus, in which my name is associated with that of Professor Sylvester, without expressing my deep gratitude for the kind interest which he took in my researches, and my regret that his departure for America to undertake the post of Professor in the new Johns Hopkins University has deprived me of one whose valuable suggestions and encouragement helped me much in my investigations.

Before leaving the Peaucellier cell and its modifications, I must point out another important property they possess besides that of furnishing us with exact rectilinear motion. We have seen that our simplest linkwork enables us to describe a circle of any radius, and if we wished to describe one of ten miles' radius the proper course would be to have a ten-mile link, but as that would be, to say the least, cumbrous, it is satisfactory to know that we can effect our purpose with a much smaller apparatus. When the Peaucellier cell is mounted for the purpose of describing a straight line, as I told you, the distance between the fixed pivots must be the same as the length of the "extra" link. If this distance be not the same we shall not get straight lines described by the pencil, but circles. If the difference be slight the circles described will be of enormous magnitude, decreasing in size as the difference increases. If the distance QO, Fig. 6, be made greater than QC, the convexity of the portion of the circle described by the pencil (for if the circles are large it will of course be only a portion which is described) will be towards O, if less the concavity. To a mathematician, who knows that the inverse of a circle is a circle, this will be clear, but it may not be amiss to give here a short proof of the proposition.

In Fig. 17 let the centers Q, Q' of the two circles be at distances from O proportional to the radii of the circles. If then $ODCPS$ be any straight

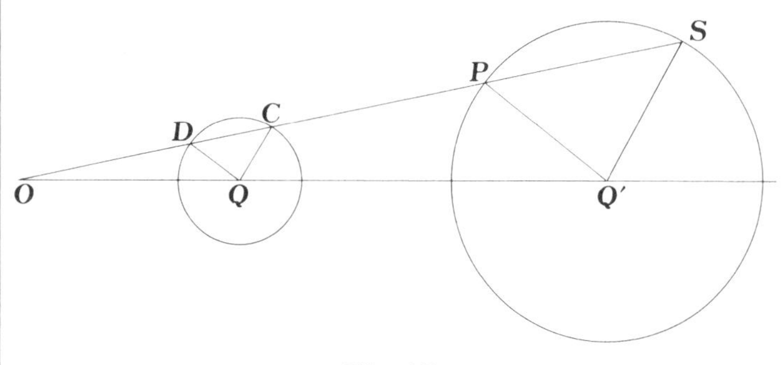

Fig. 17.

line through O, DQ will be parallel to PQ', and CQ to SQ', and OD will bear the same proportion to OP that OQ does to OQ'. Now considering the proof we gave in connection with Fig. 7, it will be clear that the product $OD \cdot OC$ is constant, and therefore since OP bears a constant ratio to OD,

$OP \cdot OC$ is constant. That is if $OC \cdot OP$ is constant and C describes a circle about Q, P will describe one about Q'. Taking then O, C and P as the O, C and P of the Peaucellier cell in Fig. 7, we see how P comes to describe a circle.

It is hardly necessary for me to state the importance of the Peaucellier compass in the mechanical arts for drawing circles of large radius. Of course the various modifications of the "cell" I have described may all be employed for the purpose. The models exhibited in the Conservatoire by M. Breguet are furnished with sliding pivots for the purpose of varying the distance between O and Q, and thus getting circles of any radius.

My attention was first called to these linkworks by the lecture of Professor Sylvester, to which I have referred. A passage in that lecture in which it was stated that there were probably other forms of seven-link parallel motions besides M. Peaucellier's, then the only one known, led me to investigate the subject, and I succeeded in obtaining some new parallel motions of an entirely different character to that of M. Peaucellier (9). I shall bring two of these to your notice, as the investigation of them will lead us to consider some other linkworks of importance.

If I take two kites, one twice as big as the

other, such that the long links of each are twice the length of the short ones, and make one long link of the small kite lie on a short one of the large, and a short one of the small on a long one of the large, and then amalgamate the coincident links, I shall get the linkage shown in Fig. 18.

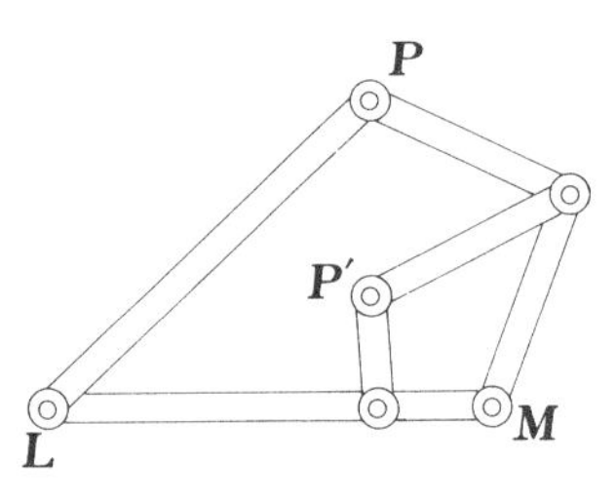

Fig. 18.

The important property of this linkage is that, although we can by moving the links about, make the points P and P' approach to or recede from each other, the imaginary line joining them is always perpendicular to that drawn through the pivots on the bottom link LM. It follows that if either of the pivots P or P' be fixed, and the link LM be made to move so as always to remain parallel to a fixed line, the other point will describe a straight line perpendicular to the fixed line. Fig. 19 shows you the parallel motion made by fixing P'. It is unnecessary for me to point out how the parallelism of LM is preserved by adding the link

SL, it is obvious from the figure. The straight line which is described by the point *P* is perpendicular to the line joining the two fixed pivots; we can, however, without increasing the number of links,

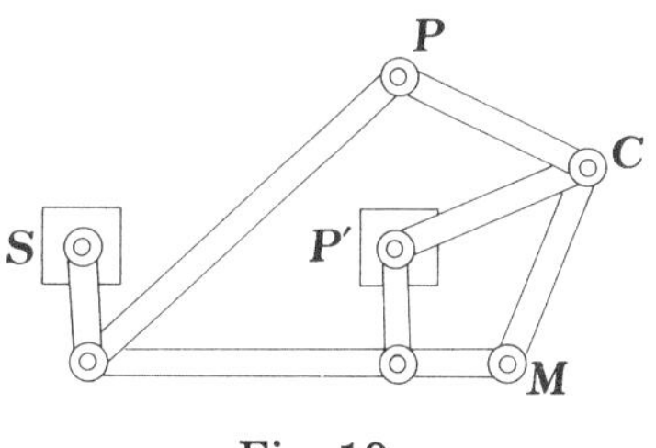

Fig. 19.

make a point on the linkwork describe a straight line inclined to the line *SP* at any angle, or rather we can, by substituting for the straight link *PC* a plane piece, get a number of points on that piece moving in every direction.

In Fig. 20, for simplicity, only the link *CP′* and the new piece substituted for the link *PC* are shown. The new piece is circular and has holes pierced in it all at the same distance—the same as the lengths *PC* and *P′C*—from *C*. Now we have seen from Fig. 19 the *P* moves in a vertical straight line, the distance *PC* in Fig. 20 being the same as it was in Fig. 19; but from a well-known property of a circle, if *H* be any one of the holes pierced in the piece, the angle *HP′P* is constant, thus the straight line HP' is fixed in position, and H moves along it; similarly all the other holes move along in straight lines passing through the fixed pivot P', and we get straight line motion distributed in all directions. This species of motion is called by Professor Sylvester "tram-motion." It is worth noticing that the motion of the circular disc is the

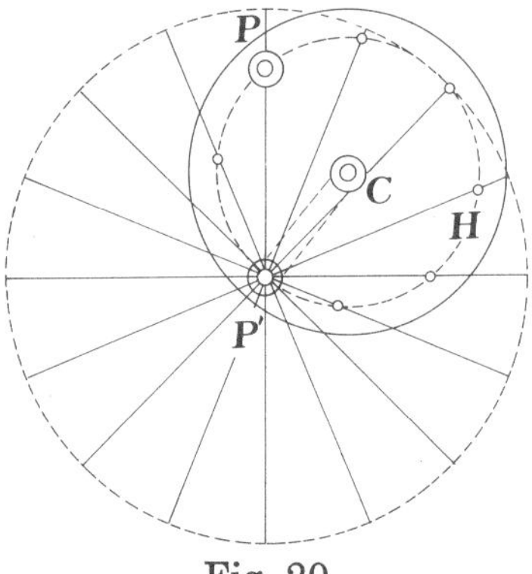

Fig. 20.

same as it would have been if the dotted circle on it rolled inside the large dotted circle; we have, in fact, White's parallel motion reproduced by linkwork. Of course, if we only require motion in one direction, we may cut away all the disc except a portion forming a bent arm containing C, P, and the point which moves in the required direction.

The double kite of Fig. 18 may be employed to form some other useful linkworks. It is often necessary to have, not a single point, but a whole piece moving so that all points on it move in straight lines. I may instance the slide rests in lathes, traversing tables, punches, drills, drawbridges, etc. The double kite enables us to produce linkworks having this property. In the linkwork of Fig. 21, the construction of which will be at once

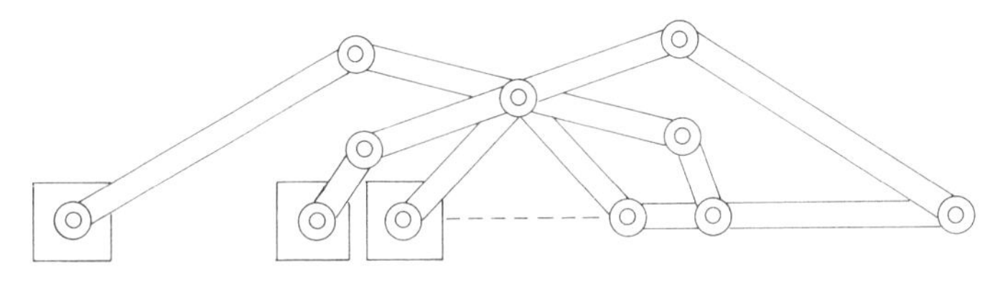

Fig. 21.

appreciated if you understand the double kite, the horizontal link moves to and fro as if sliding in a fixed horizontal straight tube. This form would possibly be useful as a girder for a drawbridge.

In the linkwork of Fig. 22, which is another combination of two double kites, the vertical link moves so that all its points move in horizontal straight lines. There is a modification of this linkwork which will, I think, be found interesting. In the linkage in Fig. 23, which, if the thin links are removed, is a skeleton drawing of Fig. 22, let the dotted links be taken away and the thin ones be inserted; we then get a linkage which has the same

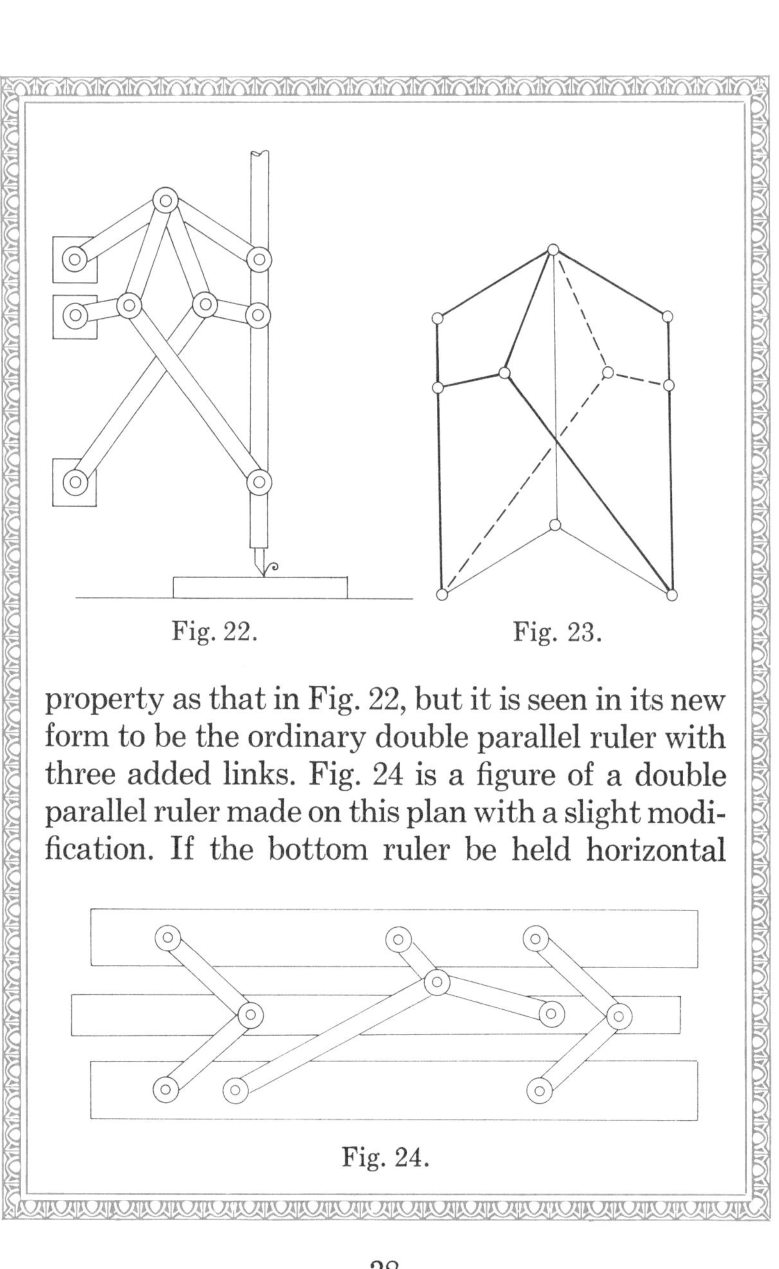

Fig. 22.

Fig. 23.

property as that in Fig. 22, but it is seen in its new form to be the ordinary double parallel ruler with three added links. Fig. 24 is a figure of a double parallel ruler made on this plan with a slight modification. If the bottom ruler be held horizontal

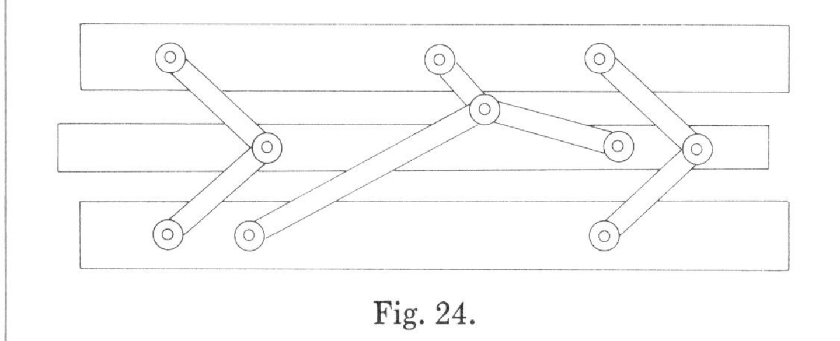

Fig. 24.

the top moves vertically up and down the board, having no lateral movement.

While I am upon this sort of movement I may point out an apparatus exhibited in the Loan Collection by Professor Tchebicheff, which bears a strong likeness to a complicated camp-stool, the seat of which has horizontal motion. The motion

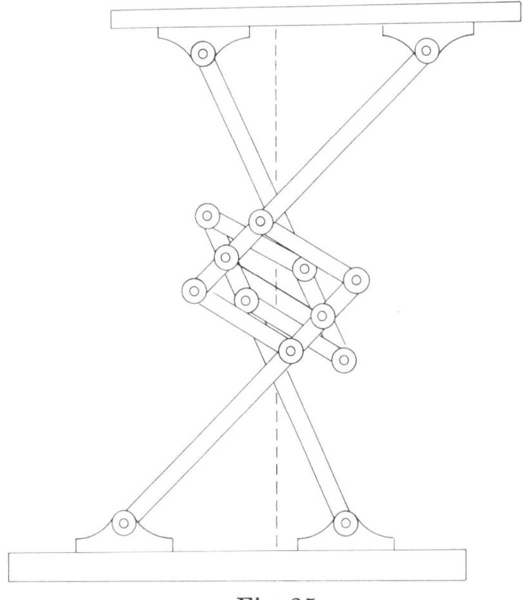

Fig. 25.

is not strictly rectilinear; the apparatus being—as will be seen by observing that the thin line in the figure is of invariable length, and a link might therefore be put where it is—a combination of two of the parallel motions of Professor Tchebicheff given in Fig. 4, with some links added to keep the seat parallel with the base. The variation of the upper plane from a strictly horizontal movement is therefore double that of the tracer in the simple parallel motion.

Fig. 26 shows how a similar apparatus of much simpler construction, employing the Tchebicheff approximate parallel motion can be made. The lengths of the links forming the parallel motion have been given before (Fig. 4). The distance between the pivots on the moving seat is half that between the fixed pivots, and the length of the remaining link is one-half that of the radial links.

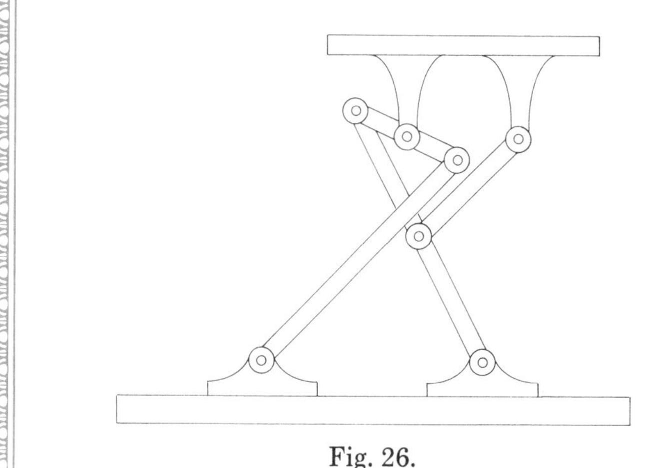

Fig. 26.

An *exact* motion of the same description is shown in Fig. 27. O, C, O', P are the four *foci* of the quadriplane shown in the figure in which the links are bent through a right angle, so that $OC \cdot OP$ is constant, and COP a right angle. The focus O is pivoted to a fixed point, and C is made by

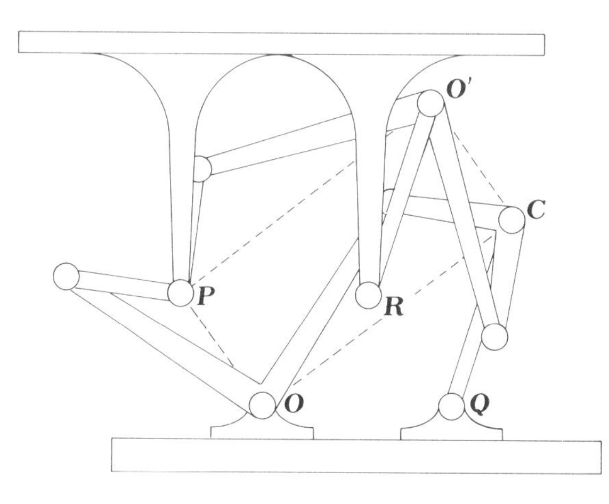

Fig. 27.

means of the extra link QC to move in a circle of which the radius QC is equal to the pivot distance OQ. P consequently moves in a straight line parallel to OQ, the five moving pieces thus far described constituting the Sylvester-Kempe parallel motion. To this are added the moving seat and the remaining link RO', the pivot distances of which,

PR and RO', are equal to OQ. The seat in consequence always remains parallel to QO, and as P moves accurately in a horizontal straight line, every point on it will do so also. This apparatus might be used with advantage where a very smoothly-working traversing table is required.

I now come to the second of the parallel motions I said I would show you. If I take a kite and pivot the blunt end to the fixed base, and make the sharp end move up and down in a straight line, passing through the fixed pivot, the short links will rotate about the fixed pivot with equal velocities in opposite directions; and, conversely, if the links rotate with equal velocity in opposite directions, the path of the sharp end will be a

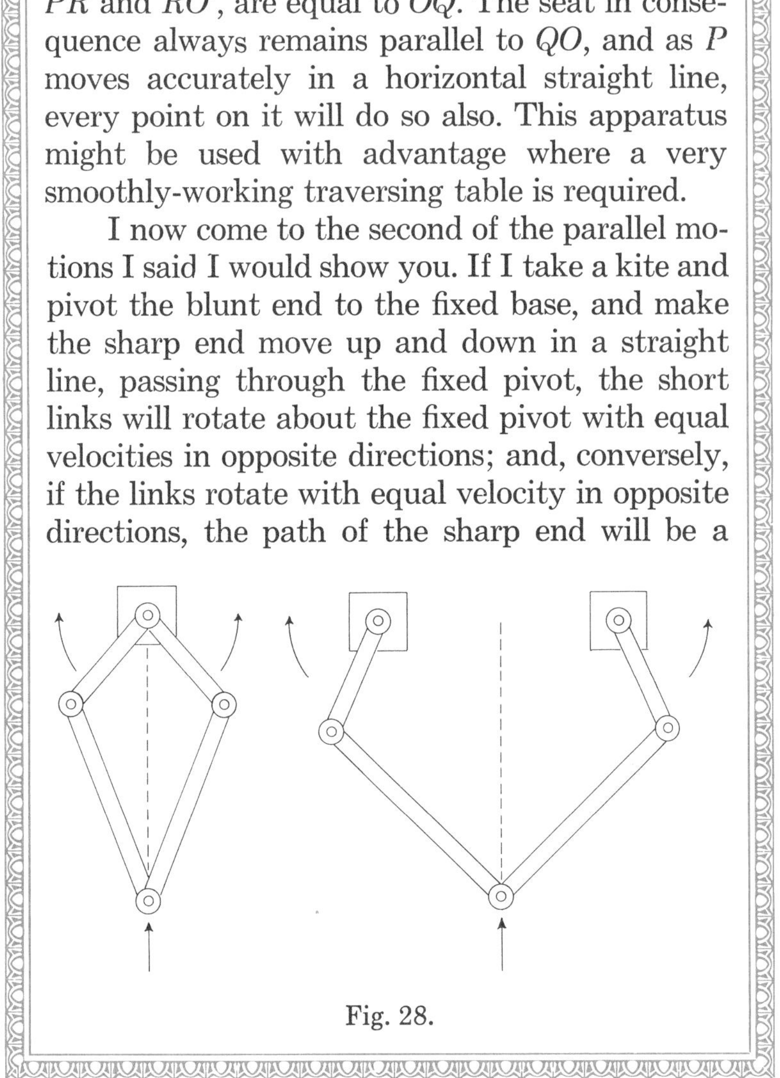

Fig. 28.

straight line, and the same will hold good if instead of the short links being pivoted to the same point they are pivoted to different ones.

To find a linkwork which should make two links rotate with equal velocities in opposite directions was one of the first problems I set myself to solve. There was no difficulty in making two links rotate with equal velocities in the same direction,—the ordinary parallelogrammatic linkwork employed in locomotive engines, composed of the engine, the two cranks, and the connecting

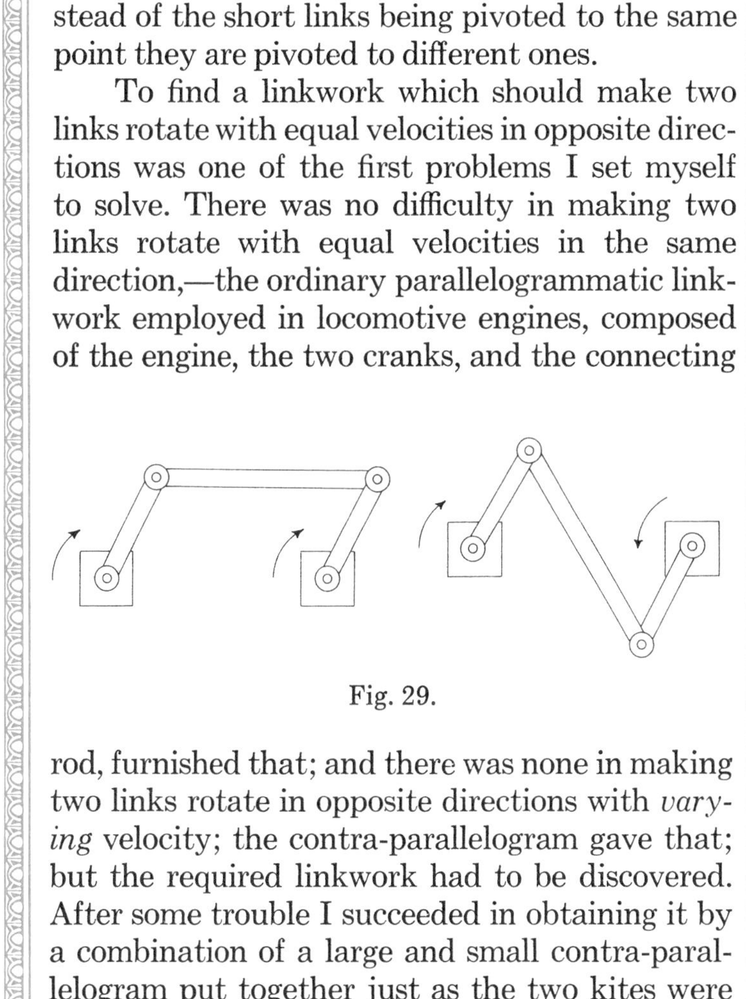

Fig. 29.

rod, furnished that; and there was none in making two links rotate in opposite directions with *varying* velocity; the contra-parallelogram gave that; but the required linkwork had to be discovered. After some trouble I succeeded in obtaining it by a combination of a large and small contra-parallelogram put together just as the two kites were in the linkage of Fig. 18. One contra-parallelogram is made twice as large as the other, and the long links of each are twice as long as the short (10).

The linkworks in Figs. 30 and 31, will, by considering the thin line drawn through the fixed pivots in each as a link, be seen to be formed by fixing different links of the same six-link linkage composed of two contra-parallelograms as just stated. The pointed links rotate with equal velocity in opposite directions, and thus, as shown in Fig. 28, at once give parallel motions. They can of course, however, be usefully employed for the mere purpose of reversing angular velocity (11).

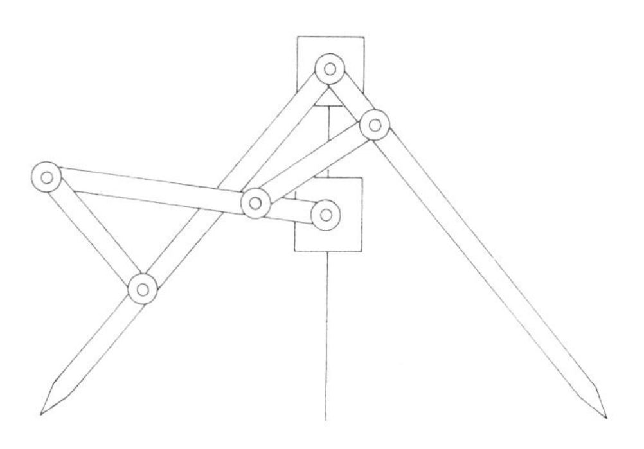

Fig. 30.

An extension of the linkage employed in these two last figures gives us an apparatus of considerable interest. If I take another linkage contra-parallelogram of half the size of the smaller one and fit it to the smaller exactly as I fitted the smaller to the larger, I get the eight-linkage of

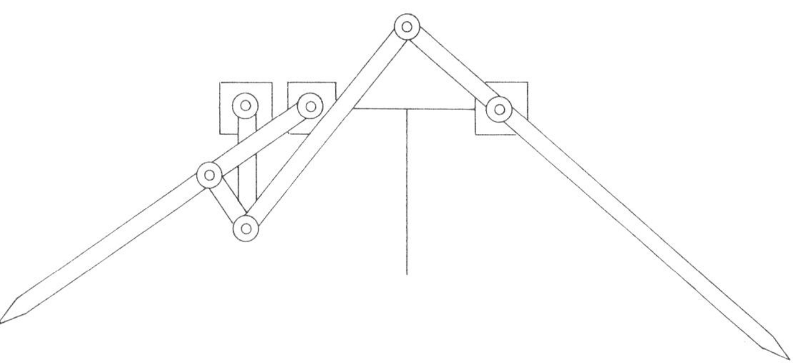

Fig. 31.

Fig. 32. It has, you see, four pointed links radiating from a center at equal angles; if I open out the two extreme ones to any desired angle, you will see that the two intermediate ones will exactly *trisect the angle.* Thus the power we have had to call into operation in order to effect Euclid's first Postulate—linkages—enables us to solve a problem which has no "geometrical" solution. I could of course go on extending my linkage and get others which would divide an angle into any number of equal parts. It is obvious that these same linkages can also be employed as linkworks for doubling, trebling, etc., angular velocity (12).

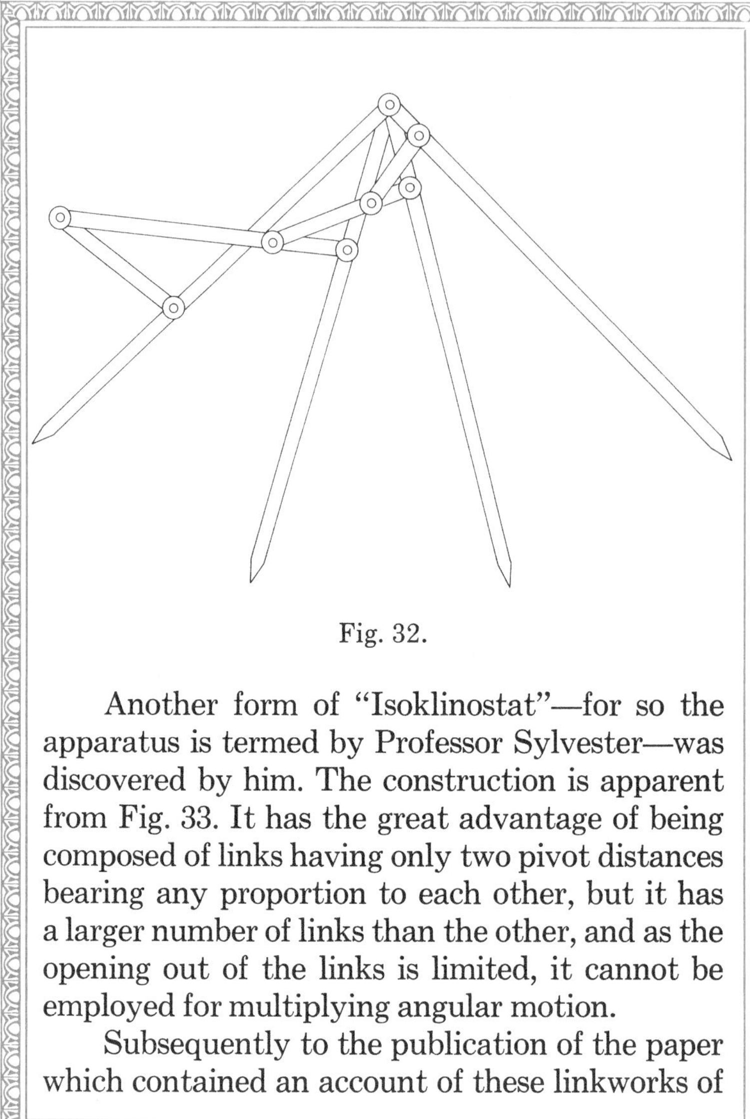

Fig. 32.

Another form of "Isoklinostat"—for so the apparatus is termed by Professor Sylvester—was discovered by him. The construction is apparent from Fig. 33. It has the great advantage of being composed of links having only two pivot distances bearing any proportion to each other, but it has a larger number of links than the other, and as the opening out of the links is limited, it cannot be employed for multiplying angular motion.

Subsequently to the publication of the paper which contained an account of these linkworks of mine of which I have been speaking, I pointed out in a paper read before the Royal Society (13) that the parallel motions given there were, as well as those of M. Peaucellier and Mr. Hart, all particular cases of linkworks of a very general character, all of which depended on the employment

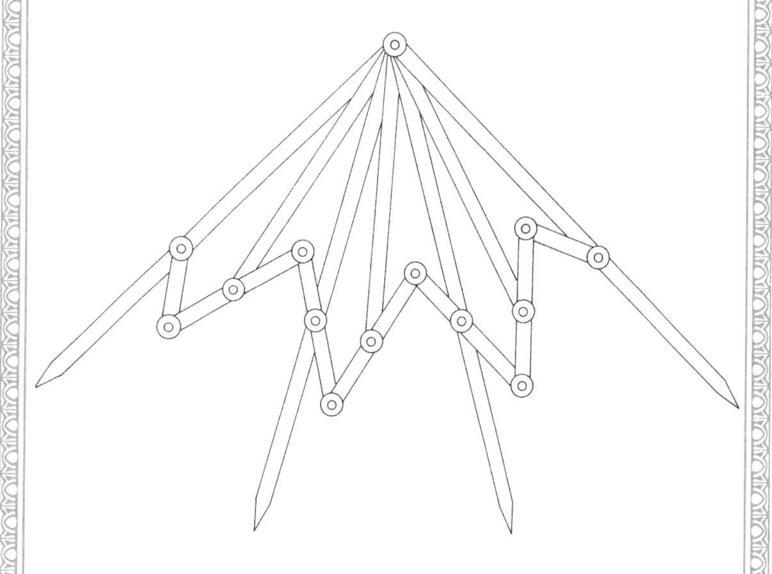

Fig. 33.

of a linkage composed of two similar figures. I have not sufficient time, and I think the subject would not be sufficiently inviting on account of its mathematical character, to dwell on it here, so I will leave those in whom an interest in the question has been excited to consider the original paper.

At this point the problem of the production of straight-line motion now stands, and I think you will be of opinion that we hardly, for practical purposes, want to go much farther into the theoretical part of the question. The results that have been obtained must now be left to the mechanician to deal with, if they are of any practical value. I have, as far as what I have undertaken to bring before you today is concerned, come to the end of my tether. I have shown you that we *can* describe a straight line, and *how* we can, and the consideration of the problem has led us to investigate some important pieces of apparatus. But I hope that this is not all. I hope that I have shown you (and your attention makes that hope a belief) that this new field of investigation is one possessing great interest and importance. Mathematicians have no doubt done much more than I have been able to show you today (14), but the unexplored fields are still vast, and the earnest investigator can hardly fail to make new discoveries. I hope therefore that you whose duty it is to extend the domain of science will not let the subject drop with the close of my lecture.

NOTES

(1) The hole through which the pencil passes can be made to describe a circle independently of any surface (see the latter part of Note 3), but when we wish to describe a circle on a given plane surface that surface must of course be assumed to be plane.

(2) "But" (it is carefully added) "not a graduated one." By the use of a ruler with only two graduations, an angle can, as is well known, be readily trisected, thus —Let RST be the angle, and let PP' be the points where the graduations cut the edge of the ruler. Let $2RS = PP'$. Draw RU parallel and RV perpendicular to ST. Then if we fit the ruler to the figure $RSTUV$ so that the edge PP' passes through S, P lies on RU and P' on RV, PP' trisects the angle RST. For if Q be the middle point of PP', and RQ be joined, the angle TSP = the angle QRP = the angle QRP = half the angle RQS, that is half the angle RSQ.

This solution is of course not a "geometrical" one in the sense I have indicated, because a graduated ruler and the fitting process are employed. But does Euclid confine himself to his three Postulates of construction? Does he not use a graduated ruler and this fitting process? Is not the side AB of the triangle ABC in Book I. Proposition 4, graduated at A and B, and are we not told to take it up and fit it on to DE?

It seems difficult to see why Euclid employed the second Postulate—that which requires "that a terminated straight line may be produced to any length in a straight line,"—or rather, why he did not put it among the propositions in the First Book as a problem. It is by no means difficult by a rigid adherence to Euclid's methods to find a point outside a terminated straight line which is in the same straight line with it, and to prove it to be so, without the employment of the second Postulate. That point can then, by the first Postulate, be joined to the extremity of the given straight line which is thus produced, and the process can be continued indefinitely, since by the third Postulate circles can be drawn with any center and radius.

(3) It is important to notice that the fixed base to which the pivots are attached is really one more link in the system. It would on that account be perhaps more scientific, in a general consideration of the subject, to commence by calling any combination of pieces (whether those pieces be cranks, beams, connecting-rods, or anything else) jointed or pivoted together, a "*linkage*." When the motion of the links is confined to one plane or to a number of parallel planes, the system is called a "*plane linkage*." (It will be seen that this lecture is confined to plane linkages; a few remarks about solid linkages will be found at the end of the note.) The motion of the links among themselves in a linkage may be determinate or not. When the motion is determinate the number of links must be even, and the linkage is said to be "*complete*." When the motion is not determinate the linkage is said to have 1, 2, 3, etc. degrees of freedom, according to the amount of liberty the links possess in their relative

motion. These linkages may be termed "*primary*," "*secondary*," etc. linkages. Thus if we take the linkage composed of four links with two pivots on each, the motion of each link as regards the others is determinate, and the linkage is a "*complete linkage*." If one link be jointed in the middle the linkage has one degree of liberty and is a "*primary linkage*." So by making fresh joints "*secondary*" or "*tertiary*," etc. linkages may be formed. These primary, etc. linkages may be formed in various other ways, but the example given will illustrate the reason for the nomenclature. When one link of a linkage is a fixed base the structure is called a "*linkwork*." Linkworks, like linkages, may be "*primary*," "*secondary*," etc. A "*complete linkwork*," *i.e.* one in which the motion of every point on the moving part of the structure is definite, is called a "*link-motion*." The various "grams" described by these link-motions are very difficult to deal with. I have shown, in a paper in the *Proceedings of the London Mathematical Society*, 1876, that a link-motion can be found to describe any given algebraic curve, but the converse problem, "Given the link-motion, what is the curve?" is one towards the solution of which but little way has been made; and the "tricircular trinodal sextics," which are the "grams" of the simple three-piece motion, are still under the consideration of some of our most eminent mathematicians.

Taking them in their greatest generality, the theoretically simplest form of link-motion is not the flat circle-producing link, but a solid link pivoted to a fixed center, and capable of motion in all directions about the center, so that all points on it describe spheres in space; and the most general form a number of such links pivoted together, forming a structure the various points on which describe surfaces. If two simple solid links, turning about two fixed centers, are pivoted together at a common point, that point will describe a circle independently of any plane surface, the other points on the links describing portions of spheres. The form of pivot which would have to be adopted in solid linkages would be the ball-and-socket joint, so that the links could not only move about round the fixed center, but rotate about any imaginary axis through that center. It is obvious that it would be impossible to construct any joint which would give the links perfect freedom of motion, as the fixed center about which any link turned must be fastened to a fixed base in some way, and whatever means were adopted would interfere with the link in some portion of its path. This is not so in plane link-motions. The subject of solid linkages has been but little considered. Hooke's joint may be mentioned as an example of a solid link-motion. (See also Note 11.)

(4) I have been more than once asked to try and get rid of the objectionable term "parallel motion." I do not know how it came to be employed, and it certainly does not express what is intended. The apparatus does not give "parallel motion," but approximate "rectilinear motion." The expression, however, has now become crystallized, and I for one cannot undertake to find a solvent.

(5) See the *Proceedings of the Royal Institution,* 1874.

(6) This paper is printed *in extenso* in the *Cambridge Messenger of Mathematics,* 1875, vol. iv., pp. 82-116, and contains much valuable matter about the mathematical part of the subject.

(7) The interchange of a radial and traversing bar converts Watt's Parallel Motion into the Grasshopper Parallel Motion. The same change shows us that the curves traced by the linkwork formed by fixing one bar of a "kite" are the same as those traced by the linkwork formed by fixing one bar of a contra-parallelogram. This is interesting as showing that there is really only one case in which the sextic curve, the "gram" of three-bar motion, breaks up into a circle and a quartic.

(8) For a full account of this and the piece of apparatus next described, see *Nature,* vol. xii., pp. 168 and 214.

(9) See the *Messenger of Mathematics,* "On Some New Linkages," 1875, vol. iv., p. 121.

(10) A reference to the paper referred to in the last note will show that it is not necessary that the small kites and contra-parallelograms should be half the size of the large ones, or that the long links should be double the short; the particular lengths are chosen for ease of description in lecturing.

(11) By an arrangement of Hooke's joints, pure solid linkages, we can make two axes rotate with equal velocities in contrary directions (See Willis's *Principles of Mechanism,* 2nd Ed. sec. 516, p. 456), and therefore produce an exact parallel motion.

(12) The "kite" and the "contra-parallelogram" are subject to the inconvenience (mathematically very important) of having "dead points." These can be, however, readily got rid of by employing pins and gabs in the manner pointed out by Professor Reuleaux. (See Reu-

leaux's *Kinematics of Machinery*, translated by Professor Kennedy, Macmillan, pp. 290-294.)

(13) *Proceedings of the Royal Society*, No. 163, 1875, "On a General Method of Obtaining Exact Rectilinear Motion by Linkwork." I take this opportunity of pointing out that the results there arrived at may be greatly extended from the following simple consideration.

If the straight link *OB* makes any angle *D* with the straight link *OA*, and if instead of employing the straight links we employ the pieces *A'OA*, *B'OB*, and if the angle

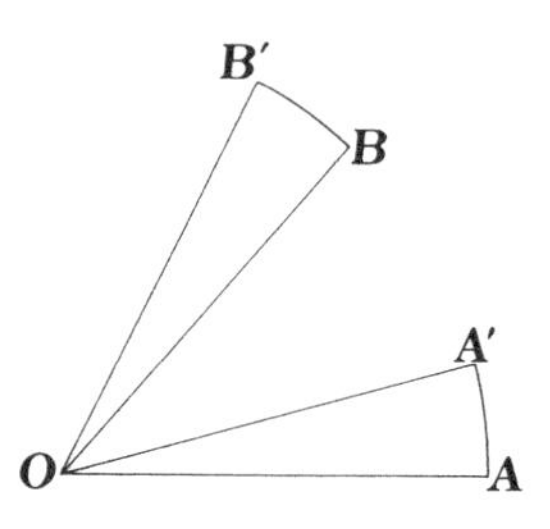

Fig. 34.

A'OA equals the angle *B'OB*, then the angle *B'OA'* equals *D*. The recognition of this very obvious fact will enable us to derive the Sylvester-Kempe parallel motion from that of Mr. Hart.

(14) In addition to the authorities already mentioned, the following may be referred to by those who desire to know more about the mathematical part of the subject of "Linkages." "*Sur les Systèmes de Tiges Articulées,*" par M. V. Liguine, in the *Nouvelles Annales*, December, 1875, pp. 520-560.

Two papers "*On Three-bar Motion,*" by Professor Cayley and Mr. S. Roberts, in the *Proceedings of the London Mathematical Society*, 1876, vol. vii., pp. 14 and 136. Other short papers in the *Proceedings of the London Mathematical Society*, vols. v., vi., vii., and the *Messenger of Mathematics*, vols. iv. and v.